This Science Notebook Belongs To:

Name:

Contact Information:

Sir Isaac Newton

Sir Isaac Newton

Isaac Newton was born in Woolsthorpe, Lincolnshire, England, on December 25, 1642. He died in London, England, on March 20, 1727.

Newton was an English physicist and mathematician who represented the Scientific Revolution of the 17th century.

In optics, his discovery of how white light is made added color phenomena to the study of light and laid the groundwork for modern physical optics.

In mechanics, his three laws of motion, which are the basic building blocks of modern physics, led to the law of universal gravitation.

He was the first person to find out about the infinitesimal calculus in math.

The Philosophiae Naturalis Principia Mathematica (1687) by Isaac Newton was one of the most important books in the history of modern science.

(Isaac Newton | Biography, Facts, Discoveries, Laws, & Inventions | Britannica, 2022)

Made in the USA
Coppell, TX
04 November 2026

11286707R00057